Storm Mountain had as many legends as it had trees. According to the Indians, it was where all the rain came from. Stranger still was the legend that a mad scientist lived on the summit.

But Taylor was a sensible young man who liked to discover things on his own. So one sunny morning he decided to climb to the top of Storm Mountain and just see for himself.

Taylor packed a snack, his compass, and a raincoat, just in case the legend was true. He was all set to go when he ran into his sister Katie. She never missed a thing.

"Where are you going?" asked Katie.

"Nowhere," responded her brother.

"I bet you're going up to Storm Mountain," said Katie. "I saw you looking through the telescope."

"So what?" said Taylor.

"I bet you're going to look for the mad scientist," said Katie. "Can I go?"

"No girls," said Taylor.

"I'm not just a girl," said Katie. "I'm your sister."

As usual, Katie was very hard to argue with.

Water Watch

Written by
Patrick Smith Kelly

Illustrated by
Chris Jones and Sascha Immerman

Art Direction
Eddie Terrill

Computer Typesetting Consultant
Mike Scearce

Special Thanks to:
John M. Akey
Former President of the National Science Teachers Association
(NSTA)
Former Science Consultant for Colorado Department of Education

Jerry Emanuelson
President of Colorado Futurescience

Thanks to Roger, Shanlee, Kyle and Maynard the dog.

ISBN 0-939251-11-6

Off they went, past the edge of town, up through the meadow, around Storm Reservoir, and into the forest. The trail was steep and the sun was hot.

Taylor and Katie couldn't believe their eyes. Sitting smack on the top of Storm Mountain was a crazy little house covered with weather instruments.

And sure enough, just like the legend said, there was a mad scientist in a white lab coat. He was talking to a dog. A basset hound, to be precise.

"Watch it," whispered Taylor to his sister, "he's probably a nut case."

"I don't care," said Katie, "I'm thirsty."

"Excuse me," said Katie, "could I please have some water?"

"H-two-oh!" he shouted. "You're in luck, young lady."

"Who in the world are you?" said Katie.

"I'm Professor Weather," said the man, "and this is my assistant Stormy."

"Hello," barked the dog.

"And this is our friend Wally Water," said the Professor. "He dropped in during the rain last night."

"You wouldn't know it by looking at him," said the Professor, "but Wally is over four billion years old."

"That's funny," said Wally,"I don't feel a century over three billion."

"That's because no new water is ever made," explained the Professor. "The water we use today has fallen billions of times over millions of years."

"Wow!" screamed Wally. "The sun is giving me lots of energy. I'm beginning to feel a little vaporous."

"That's my Wally," said the Professor, "always changing but always the same."

"Are you saying that the water in my shower is the same water some dinosaur took a bath in?" asked Taylor.

"That reminds me of the time I was part of a rainstorm about seventy million years ago," said Wally. "I landed right in the eye of the biggest Tyrannosaurus Rex you've ever seen. Boy was he mad!"

"Oh no," barked Stormy. "Not another dinosaur story."

"Wally comes in three forms: liquid, gas, and solid," said the Professor, "but it's always the same old Wally."

"Like rain, fog, and snow?" asked Taylor.

"Exactly!" said the Professor. "When it's below 32 degrees Fahrenheit, Wally moves so slowly he becomes ice. But heat from the sun gets him excited."

"Excited about what?" asked Taylor.

"Evaporation," smiled the Professor. "That's when water molecules move so fast they split apart and become a gas."

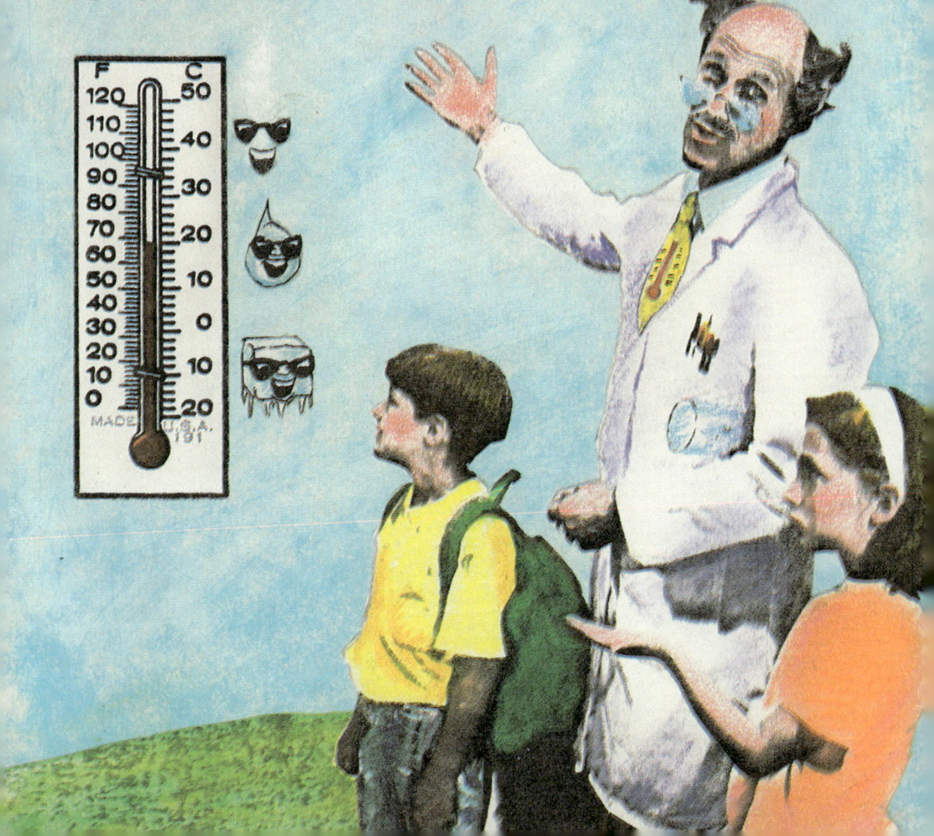

"Gotta run," said Wally, "catch you guys on the precipitation side!"

"Where's he going?" asked Katie.

"To join his friends who are evaporating from the oceans, rivers, lakes, and puddles around the world," answered the Professor. "They're all rising into the atmosphere to form clouds."

"When will Wally be back?" asked Katie.

"At the other end of the hydrologic cycle, of course," said the Professor. "Wally will eventually cool off up in the atmosphere and come falling back to earth."

"I want to go with Wally!" Katie yelled.

"Why not!" said Professor Weather. "Stormy, prepare the balloon!"

"Hurry!" yelled Katie. "He's getting away!"
"Prepare for liftoff!" barked Stormy.
"Where does a basset hound learn to fly a hot-air balloon?" asked Taylor.

"The higher we go, the colder it gets," said Professor Weather.

"The water vapor is changing back into droplets," observed Taylor.

"That's called condensation," explained the Professor. "Now the droplets will join together and fall back to earth as rain or snow, depending on the temperature of the air where the water condenses."

"That's precipitation," said Taylor.

"Precisely," said the Professor.

"Geronimo!" yelled Wally.
"Where is Wally going?" asked Katie.
"He's raining again," said the Professor.
"Parachutes anyone?" barked Stormy.
"This dog thinks of everything," said Taylor.

"Note the temperature change as we descend," shouted the Professor.
"It's getting warmer," said Taylor.
"Precisely!" said the Professor.
"Hit the silk!" barked Stormy.
"Where does a basset hound learn how to skydive?" asked Taylor.

"Note the erosion taking place," said the Professor.

"You mean the way the water is washing the sticks and soil down the mountain?" asked Taylor.

"Precisely," answered the Professor. "As the water moves, it likes to pick things up and carry them along, even rocks. Water also likes to pick up things that have been left behind by careless people, including toxic waste and chemicals."

"That's why we have to keep the mountain as clean as a whistle," barked Stormy.

"Have any of you seen Wally?" Katie asked the water droplets rushing down the mountain.

"You just missed him," said one of the rushing droplets. "He was on his way to Storm Reservoir."

"That's our Wally," barked Stormy, "he never sits still."

"Look how muddy it is," said Taylor.

"Of course," said the Professor, "there are always a lot of impurities in the water after a rain. That's why it has to be purified before we can drink it."

"Have you seen Wally?" Katie asked a droplet on the surface of the reservoir.

"He said he felt dirty," answered the droplet. "He was on his way to the water purification system for a bath."

"Untreated water is stored in the reservoir until needed," explained the Professor, "then sent into the mixing basin where it's mixed with alum, a clumping substance."

"I bet it clumps with the dirt," said Taylor.

"Quite right! The dirt sticks to the alum and settles to the bottom," explained the Professor, "then the clean water flows off the top and into the sand and gravel filter."

"Is it ready to drink?" asked Taylor.

"Not yet," said the Professor. "Some disease germs can even squirm through all that sand and gravel."

"You mean even crystal clear water can still have germs?" asked Taylor.

"Precisely," answered the Professor. "So we add a chemical called chlorine that kills every last germ."

"Pure water is stored in tanks," explained the Professor, "like this water tower."

"Have you seen Wally?" Katie asked a droplet floating inside the water tower.

"You just missed him," said the droplet. "He likes to go into town after a bath."

"The clean water is then piped under the streets and up through the taps in your house," said the Professor.

Katie walked over to a woman watering her grass.

"Have you seen Wally?" Katie asked.

"He just watered my lawn," said the woman. "Then he tried to tell me some crazy story about a dinosaur."

"That's our Wally," barked Stormy, "he never stops talking."

"What's that green stuff?" asked Katie.

"Pollution," answered the Professor, "and it may even contain poisonous substances

"But that stream goes to the reservoir," said Taylor.

"Precisely! That's why our water must always be protected and never polluted," said the Professor.

"We are what we drink," barked Stormy.

“Anything and everything gets washed into the water we need,” explained the Professor. “Garbage from the dump, even the fertilizer used by farmers.”

“But doesn’t it get purified before we drink it?” asked Taylor.

“Yes,” answered the Professor. “But if we keep polluting the water faster than we can purify it, someday there might not be enough clean water to go around.”

“What if Wally got polluted?” asked Katie on the verge of tears.

“He probably evaporated already,” said Taylor.

“Don’t worry about Wally,” barked Stormy, “he can take care of himself.”

“But only with our help,” warned the Professor. “Clean water is everyone’s responsibility.”

"Even air pollution ends up in the water we need," said the Professor.

"You mean smoke from this power plant?" asked Taylor.

"Precisely," answered the Professor.

"What goes up must come down," barked Stormy.

"Chemically speaking," explained the Professor, "sulfur dioxide and nitrogen oxides are created by power plants that burn fossil fuels like oil and coal."

"You mean dinosaur fossils?" asked Taylor.

"Some dinosaurs," replied the Professor, "but fossil fuels are mostly made from plants and microscopic animals that died millions of years ago."

"These pollutants rise into the clouds and mix with water vapor to produce sulfuric and nitric acid," explained the Professor. "When those clouds start to rain, the acid falls back to earth."

"That's acid rain!" said Katie.

"Precisely," said the Professor, "and acid rain is bad for everything and everybody."

"Car exhausts are another big source of air pollution and acid rain," said the Professor.

"Because they run on fossil fuels?" asked Taylor.

"Precisely," answered the Professor. "More cars mean more acid rain."

"I prefer my unicycle," barked Stormy.

"Where does a basset hound learn to ride a unicycle?" asked Taylor.

"Will acid rain hurt me if it lands on my head?" asked Katie.

"No," said the Professor. "The acid is too weak to burn your skin. But over time, acid rain damages many things."

"Umbrellas anyone?" barked Stormy.

"This dog is more prepared than an Eagle Scout," said Taylor.

"Have you seen Wally?" Katie asked an acid rain droplet.

"Put me down!" the droplet screamed. "I don't know anybody named Wally!"

"Acid rain is always incredibly rude," barked Stormy.

"Over time, acid rain can eat right through stone like that statue," explained the Professor. "It also damages the buildings we live and work in."

"Will it eat the ground away?" asked Katie.

"No," replied the Professor, "but it can ruin the soil, making it hard to grow the crops we need."

"Wally would never do a thing like that," said Katie.

"That tree looks sick," observed Taylor.

"That's because the acid rain has soaked into the ground," explained the Professor, "and the tree is absorbing the acid rain up through its roots."

"Will the tree die?" asked Katie.

"It might," said the Professor, "acid rain is killing forests all around the world. That's bad for human beings because trees help purify our water by releasing clean water vapor from their leaves."

"Acid rain also poisons our lakes and streams," said the Professor. "Fish can't live in highly acidic water."

"Have you seen Wally?" Katie asked a fish in the park pond.

"I can't see anything right now," explained the fish, "my eyes are burning from the acid rain."

"If acid rain is so bad," asked Taylor, "why can't we stop it?"

"We're trying, but not hard enough," said the Professor. "The first step is making people realize that acid rain is everyone's problem. In fact, the smoke from our power plant can cause acid rain thousands of miles away."

"That means a faraway power plant could cause acid rain right here in our town," said Taylor.

"Precisely," answered the Professor, "because the dangerous gases get picked up by the wind and blown far over the horizon."

"Why not turn off the power plant?" asked Taylor.

"Because we need electricity for our homes and schools and businesses," replied the Professor. "One solution is to put scrubbers on the power plant's smokestack. They would scrub the smoke clean of the pollutants that cause acid rain."

"Cars are a big part of our air pollution problem," said the Professor.

"Driving less and using public transportation like buses and trains would help reduce the pollutants that cause acid rain."

"All aboard!" barked Stormy.

"Where would a basset hound learn to drive a bus?" asked Taylor.

“Clean water is everyone’s responsibility,” said the Professor.

“Even kids?” asked Katie.

“Especially kids,” replied the Professor, “because you’ll need even more clean water when you grow up.”

“And so will dogs,” barked Stormy.

"The earth's population is growing," explained the Professor, "and that means we will require more and more clean water for our cities and towns."

"Don't more people mean more pollution?" asked Taylor.

"Precisely," said the Professor. "That's why it's so important to develop clean sources of energy."

"I dig solar-powered music," barked Stormy.

"Have you seen Wally?" asked Katie.

"You just missed him," answered the droplet, "he was up in the cloud with me."

"I guess I'll never see Wally again," said Katie sadly.

"Don't be so sure," barked Stormy.

"Clean water is the most important thing in the world," said the Professor. "If we all work together, we can make sure there will be enough."

"I guess the Indian legend isn't true," said Taylor. "All the rain doesn't come from the top of this mountain."

"Rain comes from everywhere," explained the Professor, "that's why pollution affects everybody everywhere."

"Oh my goodness," said the Professor, "look who's back!"
"What kept you guys?" asked Wally.
"Where have you been?" asked Katie.
"I took another spin through the hydrologic cycle," answered Wally.
"According to my calculations," barked Stormy, "that's number 52,169,803,723."
"Where does a basset hound learn to calculate like that?" asked Taylor.

"Katie, did I ever tell you about the time I was part of a snowstorm back in the Ice Age?" asked Wally.

"What happened?" asked Katie.

"Well, I landed smack on the back of the biggest woolly mammoth you've ever seen," said Wally.

"Oh no," barked Stormy, "not another mammoth story!"

The Flotation Rain Gauge

Your Water Watch Book is contained inside the Flotation Rain Gauge.

As rain collects in the container, the location of the small red ball will indicate the amount of rainfall. The small container can measure up to 2.5 inches or 10 centimeters of water. If the rainfall exceeds that amount, the excess will be caught between the inside container and the outside container. Pour the excess rain into the small container for the total rainfall measurement.

Bear in mind that rainfall from the same storm can vary greatly even over short distances. Always compare your measurements to those of the official local weather report.

Follow these guidelines when setting up your rain gauge outside:

1. Always use a level spot where trees and buildings don't interfere with the rainfall hitting the gauge. Remember, frozen rain can break your plastic Rain Gauge.
2. If desired, secure the gauge by placing screws or removable nails through the two tabs at the base. If you want the gauge to hang from a pole or fence, have an adult drill a hole through the edge of the gauge, just below the cap and opposite the face showing the red ball. Then hang the gauge from a nail.

Acid Rain

The Acid Rain Test Kit with pH tablets is used to measure the acidity of water. If the liquid measured is neutral, the pH will be below 7. The lower the pH number the more acidic the water.

Normal rain has a pH of 5.0 to 5.5. To measure the acidity of the rain collected in your gauge, follow the instructions in the Acid Rain Test Kit. To measure the acidity of snow, collect a clean sample and bring it inside to melt before making the test.

Testing the Acidity of Run-Off Water

After a rain, collect water samples in a nearby stream to determine acidity. Record the location, date, time and pH. If you get a sudden change in pH, start taking measurements more frequently and at different locations to pinpoint the source of the acidity changes.

Try taking acidity measurements each hour, for several hours after a rain. Does the pH change? Remember, acidity in the air can collect in the rain.

After a snowstorm, take pH measurements in a stream of water coming from the snow. Are you getting high acidity readings? If yes, that's because snow melt can pick up acidity from the surface of the earth or from the air.

You have to collect data and ask questions before you get answers. Team work can be very valuable in collecting information and solving problems. Sometimes when you ask questions it creates more questions and then you have to collect more data. Now you are an environmentalist!

Data Table

The Data Table is used to show the location, date, measured amount, official amount, and pH of the rainfall.

Chart the amount of rainfall received in your Flotation Rain Gauge after each storm and compare your readings with the local weather report. At the end of each month, add up the total rainfall for your location and compare that amount to the official monthly rainfall, and the average historical monthly rainfall. This information can be obtained from your local weather report.

Add the rainfall for each of the twelve months together and compare that amount to the official total as reported by your weather report. Then compare your yearly total to the average annual rainfall to see if you have had a relatively wet or dry year.

Using graph paper, compare the rainfall you measured against the officially recorded amount. Use the vertical scale (top to bottom) to measure the amount of rainfall in inches or centimeters. Use the horizontal scale (left or right) to record the date. Use two differently colored pencils to indicate the rain you measured versus the offical amount.

Over time, use the graph to help determine the rainfall pattern in your area. Is it usually more or less than the officially recorded amount? Try to figure out the reasons behind such differences.

Sample Chart

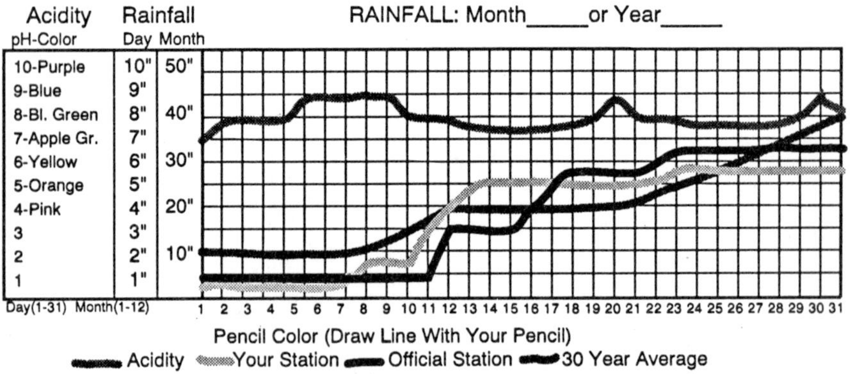